Google Glass

The Ultimate Guide for Understanding Google Glass And What You Need to Know

I0484098

presentation of the information is without contract or any type of guarantee assurance.

The trademarks that are used are without any consent, and the publication of the trademark is without permission or backing by the trademark owner. All trademarks and brands within this book are for clarifying purposes only and are the owned by the owners themselves, not affiliated with this document.

Table Of Contents

Introduction

Chapter 1: Prelude to the Revolution

Chapter 2: History of Google Glass

Chapter 3: Anatomy of Google Glass

Chapter 4: Power of Google Glass

Chapter 5: Google Glass Revolution

Chapter 6: Fear of Google Glass?

Conclusion

Introduction

This short book is for people who are interested in learning more about Google Glass and are not sure where to start or what information to rely on. I made this book in response to the high demand of people wanting an introductory book for Google Glass that breaks down why this new technology seems to be the wave of the future.

The Internet has many articles and misinformation regarding Google Glass that, unfortunately, has confused people who are interested in learning about this revolutionary craze - possibly interested in purchasing a pair for themselves. Whether you plan on using Glass yourself, or you just want to know more about why this trend is becoming so popular, it is important to first understand all the benefits and risks involved.

I recommend that you take notes while you are reading this book. This will ensure that you get the most out of the information in here. I want you to feel that you made a purchase that is worth your money and so you can look over the notes of this book even after you've finished reading it. The notes will help you to pinpoint exactly what you need to remember and by writing things down, you will be able to recall specifics if you decide to purchase.

Lastly, remember that everything in this book has been compiled through research, my own experiences, as well as the experiences of others, so feel free to question what you have read in this book. I encourage you to do your own research on the aspects that you want to look deeper into. The more you understand about Google Glass, the more educated your decision-making process will be when it comes to trying it out or giving advice to others.

Starting from the old bulky models of computers, we have witnessed everything up until the rise of smartphones and tablets. But as much as these current models of phones are useful and revolutionary, it won't be long until they become obsolete as well.

Computers are growing smaller and smaller, and the mantra of companies at Silicon Valley still prevails: "less is more." Which leads us to the question: how small and portable could computers get? Remember those gadgets you saw in sci-fi movies? Those are quickly becoming reality and Google is spearheading efforts towards the future of the computer industry.

In 2012, Google announced its plans regarding the development of a wearable computer. *Google Glass* will soon become the world's first commercially available computer eyewear. Although the blueprint for computer eyewear (or at least the idea of creating one) has been around for decades, this is the first time that a company is investing very large amounts of money into making a prototype available to the public.

This book contains information about the history of wearable computers and the fundamental principles behind them. It also explains the development of Glass, from conception to beta-testing. We will dissect Glass to its most basic components and explain what makes it unique.

We'll also dive into the various features that Glass has (so far), such as the camera abilities, third-party application compatibility and a lot more. We'll then discuss the game-changing disruptions that Google Glass has created in the education, medical and business sectors. Also, despite the positive implications that Google Glass creates, there will always be concerns about its mass production and the necessary submission of personal information. We will briefly cover that topic as well.

Overall, whether we like it or not, Silicon Valley is leaning to a completely hands-free world. Google Glass might not be the first of its kind, but it has definitely sparked a revolution.

Chapter 1:

Prelude to the Revolution

Predecessors to Google Glass

Despite what many people think, the concept of computer eyewear isn't a new phenomenon. Even in the mid-20th century, scientists were finding innovative ways to miniaturize the computational potential of a computer, in order to make it portable. The first attempt at using wearable computers to create an *augmented reality* was in 1968, when Ivan Sutherland created a head-mounted display.

The prototype used silver linings so that people could see a virtual reality on top of the physical environment in front of them. For example, Sutherland could program the display to make a dog appear when the user was surrounded by grass. This created an illusion because the object (the dog) appears around the real background (the grass). The process of mixing illusion and reality is called augmentation.

Ironically, wearable computers didn't start as a tool that aided in education or medicine. Wearable computers in the 1970s were actually used to beat roulettes. The ability of the computer to predict and calculate chaotic events such as the spinning roulette was ground-breaking because it received the same stimuli as a human did. The only difference was, it could solve complex computations in a matter of seconds, something that couldn't be done by the best of mathematicians, much less, a regular gambler.

This was only the first of many attempts to shrink a computer. At the forefront of wearable computer technology was *Steve Mann*. Mann conducted various experiments involving wearable computing. However, he conceded that with the available technology of the 1990's, the dream of a portable computer would take years to develop. Mann is most remembered for creating an article about how a wearable computer would likely function.

Fundamentals of wearable technology

If we dissect Mann's ideas, we come up with three basic principles for wearable computers: augmentation, constancy and mediation. These three concepts would become the basis for most experiments involving wearable computers.

Augmentation

Mann believed that miniature computers would complement our limited human senses. The information that we normally received was only a fraction of what we *could* muster. In fact, it has been proven that many animals have better sensory organs than humans.

However, with miniature computers, we are able to get more information from the world around us. Say you are outside and you are wondering what the weather is like, instead of looking up the temperature on your mobile device, your computer will instantly display the weather details. Furthermore, wearable technology could aid us in increasing our memory. Mann imagined eyewear that would display the profile of the person we are talking to, which would make interactions easier.

All in all, augmentation is a primary facet of miniature computers because it is in constant interaction with our reality. And out of that reality, it extrapolates information for us to use that will help us to better navigate our surroundings and interactions.

Constancy

Some people who oppose the idea of wearable computer eyewear argue that it will make people feel uneasy. However, Mann counters this by stating that the primary aim of wearable technology shouldn't be its sheer computing power, but its ability to cloak itself in the background. Simply put, wearable computers could eventually complement the human body so much so, that people would not notice its presence in the first place!

In another sense, constancy refers to, as the word implies, being constant. The wearable computer will continuously record data. Because of the constant data gathering, the computer will be able to adjust itself depending on the person's personality and habits.

Furthermore, the constancy of such technology would even lead humans to come closer to having perfect memory. Mann theorized that wearable computers could remind us of significant events whenever we receive related stimuli. For example, if you were to see roses in a store, the computer could pop up with a notification reminding you that Valentine's Day is in the near future.

Mediation

Other than the ability of a miniaturized computer to highlight a specific object and display information, it would also be able to completely *block off* unnecessary stimuli. Mann called this ability *diminished reality*.

If wearable computers ever develop this ability, then we will be able to focus more on the things we choose to. Especially in a world full of distractions, prompting our computers to ignore ads on billboards, faces of unimportant people, or even the television, will make us more productive in our work and projects.

Chapter 2:

History of Google Glass

Early development

Fast forward half a century later, Google comes into the wearable computer scene when they first introduced Google Glass to the public through a YouTube video. The debut was received well by the masses and people were amazed at the numerous possibilities of using Glass (considering that the video was shot in a first person view). The video was released in April of 2012; however, development for Google Glass began several months earlier.

The public first got a glimpse of a prototype of Glass on April 5th, 2012, when Sergey Brin, one of the co-founders of Google, wore Glass at a charity event (quite fitting that the charity was named "Foundation Fighting Blindness").

To further enhance Glass' reputation, it was highlighted in Google's annual I/O, a much awaited yearly convention. At the event held during June of 2012, athletes shared what they were seeing in real time through a Google+ Hangout. This was followed up by consistent marketing by Google all throughout 2012, from YouTube clips to online press releases.

The Explorer program

At this point, Google had created a lot of hype around the product and people were greatly anticipating the release of Glass. It wasn't a surprise that Google was receiving tens of thousands of requests from its users to get their hands on a prototype.

Fans of Glass finally got their wish fulfilled, when in February of 2013, Google announced the launching of the *Explorer* program. To receive a prototype, applicants had to post a 140-character tweet with the hashtag #IfIHadGlass, expressing their desire to become a beta tester.

It wasn't free though, as winners had to pay $1,500 just to get their own pair of Google Glass. Furthermore, they had to travel all the way to San Francisco to personally receive a prototype at a Google marketing event. Needless to say, Google still received an enormous number of participants.

Currently, there are more than 10,000 Glass beta-testers from all walks of life. Google prompts "bold and creative individuals" to still apply, for they frequently add more testers with each Google Glass update. In order to clean up bugs and add new features, Google also made the MyGlass app in order to get feedback for their product.

After testing Glass at home, Google was now ready to expand its market. People were randomly selected in the streets of the United Kingdom to give feedback about their experience with Google Glass. In Spain, Glass was even used by the Prime Minister at a political caucus.

Reaching out to medicine

Google Glass officially contributed to the medical sector, when on June 20th, 2013, Dr. Rafael Grossman live-streamed a surgery via a Google+ Hangout. The following day, a doctor from Spain performed a knee surgery along with another surgeon located in California, who collaborated using Google+.

Chapter 3:

Anatomy of Google Glass

Enough of the boring history lectures! The question that concerns consumers and critiques alike is, what makes Google Glass tick? In this chapter, we will briefly dissect the inner workings of Glass in order to discover the mechanisms behind this small but powerful machine.

The frame (and where's the lens?)

Ironically, Google Glass isn't a pair of prescription glasses at all. Google Glass is just a mini-computer with a prism screen attached to a frame, made of lightweight materials. However, Google announced that it would make Glass compatible for people with prescription lenses for approximately $225.

Battery

Google Glass is powered by a compact Lithium polymer battery attached to the right end of the frame. Reviewers peg the Glass' performance to approximately 8 hours of continuous use. Although Glass currently runs short in the power department, Google will surely invest in further miniaturizing battery packs in order to extend Glass' lifespan.

Touchpad

On the right part of the Google Glass frame is a horizontal touchpad made for navigation. When users don't want to use the voice recognition feature of Glass, they can instead opt to manually select commands by using a variety of gestures through the touchpad.

Prism Display

Attached to the touchpad is a box-shaped glass prism. This prism measures less than 2 inches; however, when the glasses are worn, the prism appears like a 25 inch TV screen from 2.4 feet away.

Needless to say, the technology used in the prism display is a marvel of technological innovation. Unlike most displays, which only rely on LED, Glass also utilizes a new display technology called Liquid Crystal on Silicon (LCoS). The illusion appears similar to a television because of the numerous reflections and refractions that light undergoes once it enters the prism.

Camera

The Explorer edition has a 5 megapixel camera capable of taking HD photos and 720p resolution videos. Although 5 megapixels wouldn't be up to par with most smartphones in the market, with many of the cameras going up to 16 megapixels, Google had to, at least for now, sacrifice camera quality in order to deliver a lighter computer.

Sound

A bone sound transducer, which is located along the ear area, enables the user to hear without other people noticing any sound. This is very appealing to users who do not want the use of their Google Glass to be obvious.

Sensors

Google Glass is equipped with five sets of sensors that adjust to the environment and the user. The accelerometer is used to measure speed by comparing the distance of an object during movements. Google primarily equipped the tool in order for the consumers to be able to use Glass as a GPS alternative. The gyroscope changes the display orientation, which is a common feature among smartphones. Glass also uses a magnetometer which serves as a compass. Because of the magnetometer, Glass can also function as a GPS.

The ambient light sensor is used along with the camera in order to take a clear picture. The sensor measures how much light is in the surroundings, then adjusts the camera's ISO and exposure values accordingly.

Moreover, a proximity sensor will help Glass approximate how near an object is to the person wearing the Glass, similar to a rear-view camera for automobiles. This sensor is not only useful for taking pictures but also for face recognition.

Connectivity

Google Glass is Wi-Fi enabled for data connection. Currently, Google is still trying to find a way for Glass to have 4G connection without relying on other sources. For now, Glass can be tethered to a smartphone via Bluetooth.

Glass inside

Of course, all of the great external hardware that Google Glass has would be useless without an efficient internal system - everything starts with the operating system. Glass runs under the version 4.0.4 of Android, a rival OS that Google pioneered against Apple's iOS.

It might not be long before Glass also rolls out the latest Google Kitkat Android version. The entire system is managed by a 1.2 GHz dual core processor with 682 MB RAM, which is on par with most smartphones. But according to Google, they might ramp up Glass' memory to 1 GB RAM in the near future.

Chapter 4:

Power of Google Glass

So far we've been talking about the history and the anatomy of Google Glass, but we haven't discussed what Glass could do with all of the fine hardware that it has. Let's take a brief look at the different features that Google Glass can bring to your life as the consumer.

Capture your precious moments

The hands-free picture and video taking ability of Glass is arguably its most competitive advantage. Any pictures taken with Glass will look as if you were looking at that moment through your own eyes. Point-of-view at its finest!

Although the current Explorer edition's camera only contains 5 megapixels, there's no doubt that Google will improve Glass' camera for the picture-savvy consumers.

A whole new definition of *connected*

Glass is deeply meshed with Google+, Google's very own social media network. Google+'s *Hangout* feature enables several people to talk in one big conversation. Video and audio can also be shared to create a more *personal* feeling. In Glass' YouTube debut, the video showed a man in a Hangout with his girlfriend on a building, overlooking the sunset. The girl replied, "it's beautiful."

Gone are the days when chat boxes reigned the social media scene. With Glass, people will not only enjoy talking to loved ones, but also see and hear their surroundings along with them!

A smartphone for your eyes (not your hands)

Remember that Google Glass runs under the Android Operating System. So can it launch applications like mobile smartphones? It definitely can! Google developed *Mirror API* in order for Glass to run local and third-party apps. The first third-party app made for Glass was from the New York Times. The app enables a user to read news in the background whenever the device notifies them of a breaking news updates.

Your travel companion

Glass has a three-axis magnetometer that serves as a compass. Combined with current GPS technology, Glass can guide you through your route with ease! Imagine an augmented map in your close peripherals while you are driving. Aside from that, you can even ask Google if it can track down how far away your friend is if you are meeting up with them.

How do we use it?

To access all of these features you have two options. One is to use Glass' voice recognition software by giving a command. Let Glass know that you are about to issue a command by saying "Ok, Glass." After that, just say your command. For example, "Ok, Glass. Take a picture of this." Once Glass recognizes the command, it will repeat what you said for verification, similar to Apple's Siri.

Another option is to manually select your command by using the touchpad located in the right part of Glass' frame. Tap the touchpad to wake Glass up and you'll be brought to the home screen which shows the date and time. You must navigate through the homepage by using a variety of gestures, much like a regular smartphone.

Chapter 5:

The Google Glass Revolution

Death to the smartphone!

Just when we thought that traditional smartphones were here to stay, it seems that hands-free products like Glass might signal their demise.

Let's compare the two gadgets when you receive a text message... When you use a phone, you'll have to get your phone from your pocket, unlock the screen and then read. With Glass, the notification pops up in the upper right corner of the screen. All you have to do is prompt Glass to open the message, either through a voice command or with the touchpad. The process of reading a text message in Glass takes roughly four seconds. "Ok Glass, read the message." Easy as that.

If all the features of the smartphone could eventually be done better by Glass, then why have a phone in the first place? This is the critical question that consumers will ask once they get their hands on Glass. Most likely, consumers will start throwing away their phones in exchange for a pair of computer eyewear, just like the trend from basic flip phones to touch-screen smart phones.

Google Glass in business

Google Glass can constantly record what a person sees and hears. How does this help businesses? Glass gives you a more detailed profile about yourself and those you interact with. Conventionally, Google crowdsources information by using its search engine.

As you may know by now, Google gets an idea of what consumers want by analyzing the search queries that people submit. Other than Google, YouTube (which is also owned by Google) is also a source for consumer preferences. Only this time, the video searches are analyzed instead of text search queries.

On the other hand, Glass enables Google not only to analyze your preferences but also your behavior. How you use Google Glass is already a wealth of information. Once Google has a general idea about what type of consumer you are, they can now relay the information to businesses.

With this information, businesses then have a better idea about the market and how to serve. They'll then be able to make new products or adjust their services in order to fit the consumer demand. In the end, you have a market telling what it wants, and businesses making products to meet those wants. It's a win-win situation because consumers get great products, while businesses earn more profit.

As if that wasn't enough incentive, Google Glass can also help businesses grow through advertisements. Ever notice how ads related to your search query appear when you use Google.com to search on the internet? Essentially, Google makes sure that each ad they display complements the user's wants. With Glass, Google can start displaying ads depending on the type of interests you have. Thus, businesses are assured that the money they invest in having their ads posted won't go to waste.

A new perspective in education

Let's use another wearable computer principle — augmentation. Eventually, wearable technology will find its way to education. It is worth mentioning that personal computers have drastically made teaching easier. The vast resources on the internet and the ability to create virtual classrooms have contributed greatly to modern pedagogy. However, personal computers still create an *impersonal atmosphere*. Ironic isn't it?

In a class with computers, you now have three actors: the teacher, the students, and the computer. Now, students will have to multi-task both listening to the teacher, and reading off the screen. As much as possible, there should be constant contact between the teacher and the students. Having a computer in between them creates unnecessary distractions. Instead, the focus should be on using computers to help integrate the student and teacher.

How does augmentation link to education? With Google Glass, all the information is fed to both the student and teacher. In the students' perspective, they'll see additional information on their screen whenever their teacher says something. This complementary information leads to a better learning experience. Students not only grasp the basics, but also gain advanced knowledge. Say the teacher is discussing World War II... While listening to the teacher, students will see visual aids such as pictures and video clips linked to WWII.

Currently, the application of Glass in education is in the film industry. In 2013, Google gave three pairs of Explorer edition glasses to three premiere art institutions, which could aid students in making first-person movies. Although Glass' contribution to the education sector isn't largely felt yet, with the use of third-party apps made possible by Glass API, Glass will become a norm in pedagogy.

Glass and the health sector

Earlier, we briefly touched on the first live-streamed surgery via Google Glass. Now, we'll discuss Glass' possible contributions to the health industry. Other than live-streaming, Glass makes it possible for surgeons to have tele-consultation with other doctors from across the world. Especially in life or death scenarios, where doctors have to make a critical decision within a matter of seconds, a quick piece of advice from another surgeon might just save a life.

Through Google+ Hangouts, medical students can witness surgeries in a first-person view. Other than learning, students can also feel as if they were the ones conducting the surgery!

Using the augmentation principle, doctors can have vital information displayed in their Glass. With constant patient information, such as heart status and blood pressure, doctors will be able to make decisions smarter and faster than ever before.

On the other hand, medical apps can aid patients in their health and/or recovery. The first medical app was created in Australia in early 2014. *Breastfeeding #ThroughGlass* helps new mothers nurse their babies. While breastfeeding, a list of instructions are displayed at the HUD. Mothers can also have consultations with a doctor through a secured Hangout if they have any concerns that the app isn't able to answer.

Chapter 6:

Fear of Google Glass?

So far, Glass seems to create a myriad of possibilities. But the truth is, not everyone is happy about those possibilities. Ever since Glass' debut, it has been subject to various criticisms. People are citing concerns about the privacy and lifestyle issues that will arise if we all live through Glass. Considering that Google is also profit-oriented, some people will undoubtedly feel insecure about being monitored by *"big brother"*.

This final chapter will serve as devil's advocate to the Google Glass argument. In the end, is Google Glass good or bad? That is something you will have to answer for yourself.

When real and virtual become one

The first idea that comes into a skeptic's mind is how Glass will blur the lines between the real and virtual world. Remember the Disney movie *Wall-E,* wherein people were so engrossed with using their gadgets that they had completely forgotten the value of personal connections?

Right now, we're slowly becoming completely reliant on technology. We use our gadgets for shopping, learning, and even interacting! According to an article in *TIME Magazine,* about the *Me, me, me Generation,* teens use text messaging so much that they even text the person beside them! Furthermore, since Facebook is so wide-reaching, many of us don't even bother meeting up with people anymore. All we do is just go through our newsfeed while liking pictures and status updates. How much more of this will occur if we all begin implementing the features of Google Glass? Will it pull the trigger to the death of our social lives?

Furthermore, the *mediation* ability of wearable technology, wherein you are able to block off other people around you, leads to a society that's unable to properly interact. In fact, critics have given a term to people who block off others — *glasshole.* The instant blocking ability of Glass will definitely deteriorate ethics in personal interactions if we are not careful.

Big brother is watching

In 2013, Edward Snowden disclosed several documents proving that the U.S. government has been perpetually monitoring its citizens. Ever since Snowden's exposé, people have become paranoid about their government's surveillance protocols. This has led to a public uproar about the violation of the right to privacy.

If we study the issue carefully, the National Security Agency is the main actor in the U.S. surveillance programs. Since the government has the mandate to take actions necessary to ensure public safety, they can resort to wire-tapping and hacking.

In fact, even Brazil's president warned the U.S. for meddling in state affairs. Moreover, Germany's chancellor lost trust in the system when she found out that the U.S. was wiretapping her phone. This only proves how surveillance issues encompass the entire world.

Once Google Glass is available to the public, there is no stopping the government from invoking its mandate to watch over our every move. It's as if we don't have personal privacy in our lives at all! Remember that Google's product will be available worldwide. And since Glass is reliant on internet connection, everything we do will be uploaded into the world data cloud.

Google might assure its users of security, but even the mighty Google is no match against the mandate of the government. And we're not just talking about the U.S. government; your identity can be compromised if black-hat hackers gain access to Google's database.

Besides that, Google will also constantly monitor your actions and behavior. How else could Google know about your preferences and personality? Earlier, we discussed how Glass will give businesses a better idea of the market. However, the question that remains is, are consumers willing to knowingly sacrifice their privacy for better products?

Safety issues

In 2013, a Californian became the first person to be ticketed for driving with Google Glass. Although the court dismissed the case due to lack of evidence, it became an eye-opener for those in the tech scene. In everyday situations like driving and walking, multi-tasking isn't a good thing. After all, there's a reason why people frown upon texting while driving. Because of Glass' accessibility, people can text, stream videos, or even take video-calls while doing something else. This puts Glass users at a risk of getting injured or even killed!

Speaking of distractions, Google might even allow ads to pop up in Glass' screen during our daily routines. When the New York Times app was rolled out, Google explicitly told the press that developers were not allowed to have ads inside the app. But, they also stated that the policy *might* change. Having ads in front of your eyes is definitely a concern, especially if the primary goal of ads is to get your attention!

The rise of new consumer policies

Did you know that even if you bought Glass for $1,500, you still don't *own it*? You read it right. Google's policy explicitly tells users that they do not have any right to sell nor loan their Glass. It might not look like much, but giving Google *partial* ownership of sold products is a sign that the government is allowing such practices. The problem arises when several companies try to emulate Google's ownership policies. Consider that Glass will likely shift the focus of companies towards wearable technology; then, tech-companies will likely have similar policies.

Conclusion

I hope this short book was able to help you learn more about Google Glass, what it is useful for, the different sectors it has been used in, and the positive and negative effects of using it. Now that you have learned about the important factors surrounding Google Glass, you can finally decide if you want to purchase it, or if you can recommend the new invention to your family and friends. Plus, a little addition to your knowledge base doesn't hurt, right? It's good to know about new innovations because it keeps you in the know and up-to-date in a world where last month's technology is already becoming outdated.

Regardless of whether Glass will become a success or a flop, there's no mistake that Google will do even more to perfect its invention. In fact, several companies are already developing their own versions of wearable computers. Google's ingenious invention is not only the dream glasses we've imagined. Rather, Google has started a race to produce the best wearable computer in the market.

Thank you for grabbing this book and I hope you learned something from it. Thanks for reading!